Time is Complicated

Time is Complicated

Kate Webb

FIERO PUBLISHING COMPANY

New York Madrid London

ISBN: 978-1-956768-32-9

Table of Contents

CHAPTER 1

"No."

"What do you mean 'no'?"

"This may come as a surprise to you, but we don't travel for frivolous reasons, Ben."

Ben raised an eyebrow. "Frivolous reasons?"

"Everyone has their own personal belief about what 'frivolous' means, but I'm the one who gets to make the final decision who goes when. You may not agree with it, however that's not going to change anything, because I am the final arbitrator."

"Yes, you are, and not many are happy with that."

"Of course they aren't. They want the freedom to do what they believe is right, without understanding how complicated this is." Peter raked a hand through his hair.

"Explaining my decisions isn't something I do often. As you're new, starting out with us, I want to talk you through why it is I make the choices I do."

"Do you believe I care?"

"Maybe not yet." Peter stood, gesturing for Ben to follow him. "The day will come when someone needs to follow in my footsteps, and it's possible that person may be you, so I want you to understand."

Following Peter through a door, into a room he wasn't sure anyone knew existed, Ben slowly looked around, seeing papers plastered all over the walls. Moving closer to one of them he was able to see handwriting all over it. The handwriting wasn't easy to understand, but from what he could read it appeared to be one of the requests that had come through for a journey.

"Time is complicated." Peter looked at Ben. "I make the decisions I do based on what I know about paradoxes, and the number of mistakes that were made in the early years. Mistakes that led to a group who have actively been working against us for the last 38 years. What I do now is try to keep people safe." He breathed in deeply. "Unfortunately keeping people safe is unpopular."

Slowly, Ben nodded. "Who are this other group?"

"According to our records they splintered off at a point when decisions were being made that they didn't agree with." Peter shrugged. "I've gone through everything from the time, wanting to understand, as well as try to mitigate..." He sat down at one of the table. "Those who chose to travel through time are rare. You're the first we've had to join us in about three years now. It's for the best, and yet it's had an effect on what we can cope with.

"During the last eight years we've lost a number of our people. Often that is through a purposely created paradox." He ran his tongue over his bottom lip. "It was created by our opposition, who want to stop us. They believe no one should be able to travel through time. Every time I let someone make a journey I put their life at risk. Sometimes it's necessary. Most of the time it's not."

"How come we haven't stopped them?"

"It's much more complicated than it appears, Ben." Peter stood once more, gesturing for Ben to follow, going over to one of the papers on the wall. "This is one of my many attempts to find a way to stop them, by going back to a point before they existed, in the hope we might be able to undo he damage before it can be done.

"Unfortunately every time I try it I create a new paradox. Trying to stop them may well cause more problems than it solves. At the same time I know I need to keep in mind how everyone views me, because I know it's entirely possible they'll make the decision to remove me. Then all I can do is hope the person who takes my position doesn't end up destroying us all."

"Can I help?"

"Probably not, as I've tried to explain this all multiple times, with most choosing not to hear what I'm saying. How dangerous can a paradox truly be?" Peter laughed. "Unfortunately far more than they realise, and there's nothing I can do to change how they see things. All I can do is protect them, considering how unwilling they are to protect themselves."

"Your words coming from someone else won't be heard either, but maybe we can find a way to stop the other group."

"Ben..."

"Let me look. Someone else looking at it may make it possible to find the answers you haven't been able to find." Their eyes met for a moment. "I know it's hard to let another person in at times, but this... I might not find anything. Dead ends might be the only thing anyone can find. If I do, though, we might be able to travel more freely than we have for a long time."

"Okay." Peter stepped over to one of the bookcases, pulling a journal out. "This will tell you some of the story."

"Some?"

"Yes, the other half belongs to the past leader of the other group, who chose not to leave his journals here. This one was written by our leader at the time, as he tried to find a way to work with them. As you know it didn't work out, because they weren't willing to hear him out, seeing him as one of the major parts of the problem, even though he did work to keep from creating any paradoxes."

"How much damage can a paradox do?"

"That's a very complicated question to answer, Ben, because there are different kinds of paradoxes. I believe they should be one of the many things a new traveller has to understand before they're free to make their first journey. On that I was overruled, as I don't control the curriculum."

"No, you don't, but I'm beginning to think you should."

Smiling, Peter moved to another bookcase. "There are three books on paradoxes here, should you wish to read them. I can understand if you don't. Each one of them is a hard read, part of the reason they've been kept off the curriculum, because we have few enough applications as it is. They don't want me to scare anyone off."

"Learning about paradoxes won't scare off anyone who's meant to be doing this." Ben joined Peter, taking the offered journal, along with the three books on paradoxes. "I think I may be beginning to understand why it was they've been trying to stop us from travelling."

"Good, then you're taking a step in the right direction. We can never become complacent, Ben, even though too many of us are. Time travel is dangerous, for both us and those we might be affecting with our choices. For a long time it was almost impossible to talk people out of going back to stop certain big things from happening, because they wanted to see how it might change the future, without understanding it might lead to their deaths.

"All it takes is one mistake to undo a future. There are multiple books that share times that have been lost, due to those choices, due to someone going through time without stopping to think for a few seconds. Instead they simply jumped in with both feet, believing they didn't need to think."

"Lost times?"

"Times you have never heard of, and neither has anyone else, other than those who actually spend time in the library. We have copies of everything there. Some know more than others, but the loudest voices, sadly, belong to those who are the most ignorant."

"Including me."

"For a time, but you've made the decision to undo that ignorance. You want to learn, which is something I am grateful for, as having someone else to argue my side may actually get some of them to listen... or at least think for a few seconds."

"Hopefully." Ben sat down at the clearest table. "I am okay to stay here, right."

"Of course. Feel free to take any of the books you may want with you, but make certain you bring them back. Certain of them are irreplaceable."

"Nothing will happen to them while I have them."

"Thank you." Peter walked toward the door. "Let me know if you need any help. Sometimes that may mean pointing you toward the right book, but I've spent a lot of time learning, in order to be the best arbitrator I can be, no matter what they might believe."

"I will."

Peter closed the door when he left the room. Ben put the journal on the table in front of him, the paradox books to one side, and took in a deep breath. No suspicion he might not be who he said he was, but then why would Peter think someone from the 'other side' would have any interest in their library, or understanding the choices that had been made in most recent times?

Removing his phone from his pocket Ben took a picture of the journal, sending it to his mentor, and then the three books on paradoxes. The phone vibrated in his hand once the second picture was sent. Answering the call, moving further away from the door, he tried to find the words he needed.

"The arbitrator is probably the best they've have, but, as he said in his own words, there are many ignorant voices working against him."

"Can we work with him?"

"Working with him isn't going to solve the problem. Most want him out, because he isn't giving them the freedom they believe they should have, and the more he tries to stop them the harder they fight against him. Putting a stop to it is going to mean bringing more of our people in than I think we wanted."

"Even with that being the case it's possibly something we need to do. Things are becoming more complicated here, as I'm sure you understand. You're safe there, for now, but there are those pushing hard to end this. If we can find some kind of balance..."

"If?" Ben shook his head. "We both know it's not going to be an easy task. They don't want to work with the travellers. They want to destroy them, believing it was the plan from the beginning, unable to understand we existed for an entirely different reason. One day we might be able to explain it to them, if they'd listen for 30 seconds. I do doubt it, but anything is possible."

"Find a way, Ben. Use what you know, what you can learn, and change things. We have enemies on both sides. If we can do it from this time then we do. If you have to travel then do it. Should we get to the point where we have no time I'll let you know."

"Good. Be careful. You know what they're going to do if they find out your helping me."

"Of course. You're the one in greater danger. All it takes is the wrong person to find out the truth."

"Don't worry about me. We both know I'm here for the right reasons."

Hanging up, phone going back into his pocket, Ben turned to find Peter was watching him, the door not quite closed behind him. "Okay, I think you need to explain who you are, and why it is you're in my library."

"As long as you're willing to have an open mind." Ben stepped over to the table. "This is the Holy Grail right now. I've read our side of the story a hundred times over, but this... it's different. This might give me the answers I need, to be able to undo the damage from before."

"Why do you want to?"

"Some of us aren't the enemy, Peter, although I will readily admit there are plenty who are. Unfortunately there are too many who still view the travellers as a danger. I don't agree, with someone like you as arbitrator. I wanted to meet you in person, which meant becoming a traveller, and asking you for an entirely frivolous journey. Then I'd know more about you. Being invited in here was the kind of luck I didn't think I'd get."

CHAPTER 2

"Let's start at the beginning. Who are you?"

Ben leaned back in his chair. "The son of the man who originally splintered off." He studied Peter. "I have no reason to think that's going to put your mind at ease, as I know you've never heard his side of the story. You have one point of view, and I have the other, which means we might have a chance to find some way to undo the damage."

"Damage?"

"Yes, damage." Ben's eyes dropped to the books on paradoxes. "We're made our share of mistakes, as much as they are those who don't want to admit it. Some of those choices have led to your people being lost." He shrugged. "It wasn't something I think my father ever wanted, from what I've read, but he's been gone a long time now. My mentor believes he's trapped in a paradox."

"The plan is to save him from that future."

"Even if it means I don't exist. I know I'm putting my own life at risk, for a man I never met, but that's who I am, unlike my brother. He blames the travellers. Had you never started journeying through time in the first place we'd still have both our parents, rather than losing one. Mom's the same. Neither of them have any idea I'm here."

"What happens if they find out?"

"Likelihood is they'll disown me." Ben looked at Peter once more. "Should I try to travel through time Alex will find a way to trap me wherever I end up, so I want to try to do this before they find out. Fortunately I have enough allies who have agreed to hide my disappearance for a time."

"Are they likely to find out?"

"It depends on how long this takes me. Knowing you haven't had any luck with finding a way to bring an end to this isn't exactly filling me with confidence."

"You could have been honest with me."

"Maybe. Trusting people is complicated, especially right now. You're someone I respect, for making the hard choices even when there are those who are fighting against you the whole time. I can't imagine how hard your task is, but I'm glad you're here, and I'm even more glad I had a chance to meet you. From what I know of my father I think he'd like you too."

"Explain to me what you know about his choices."

"Okay, to start at the beginning, you got to the point where people were time travelling for any reason they could think of, which, as you know, causes problems. There were those within the travellers who were against that, trying to get others to listen. Going anywhen at any time was only going to cause more problems. I'm certain you have some understanding of this."

Peter nodded. "I wish I could say otherwise, but I do." He ran his tongue over his bottom lip. "It still took the travellers over a decade to create the position of arbitrator, for someone like me to stop those things from happening, and ever since a loud minority has been trying to get rid of us. Unfortunately that minority has grown through time. We have the ability to travel through time, so we should be using it. At least that's what they believe."

"Your loud minority is the reason there are also a loud minority within our group who want to bring an end to the travellers entirely, to make it impossible for anyone to travel through time. They're moving toward making that happen, which is why I've been doing the opposite, because more than anything I want to help those who are trapped. If we can get them out then I'll have undone some of the worst choices my people have made. That's not something I'm willing to accept."

"The older brother will take over."

"When Mom's ready to stand down, if the time ever comes, because I know what Mom is like. She doesn't want to lose control. There have been plenty of times when I thought she was likely the one who'd trapped my father in a paradox, so she was free to take over from him. I know the only reason she went with him was because she saw an opportunity. I don't believe she actually wanted to be with him."

"Hence Alex wanting you to make a mistake."

"Of course. If he gets to take over then I think Mom would be more likely to stand down at a younger age, because she likes Alex far more than she does me. She says I'm too much like my father, which I take as a compliment, as if there's a parent I want to be like it's him. Not her."

"Yet you still don't know as much about paradoxes as you'd like."

"Unfortunately not. You still have the best books here, which is another reason I made the choice I did." Ben smiled, putting his hand on the pile. "I didn't come here to cause you problems. If I can help you I will, but my priority, right now, is helping those who can't save themselves."

"Let me help."

Ben raised an eyebrow. "You aren't kicking me out?"

"Not yet. Personally I think it would be best for the two of us to find a way to work together." Peter shrugged. "I heard you talking about possibly sending others in."

"For balance. To keep the loudest voices here from being able to take over, which would, unfortunately, have an effect on the balance back with my group, but with us here it would be easier to deal with them. We know what to expect."

"We'll see how things go. Keep an ear out, see what the situation is there, and then we'll make a decision. For now I think we're okay."

"So do I. I wouldn't ask them to come in unless there was no other choice."

"Good. Then I think we should be able to find a way." Peter stood. "For now I think I'll leave you to your reading. It may help you in your quest to understand more, as you've said that's the most important task you have in front of you right now."

As he left the room Ben watched him go. Being able to talk to Peter was definitely not something he expected, although he had hoped the time would come when they'd be able to find a way to work together, only that would be after he learnt the truth about what happened.

First he needed to understand the others involved in what happened, which meant reading the journal.

Opening the front cover, memories of doing the same thing with his father's journal coming to mind, Ben looked down at the first page. The handwriting, fortunately, was easy to read, rather than his father's messy scrawl, something he'd only been able to read due to his own handwriting. Mom was right when she said the two of them were similar.

Most of the early pages were about their time as the leader of the travellers, something they took on after their father died. Their father, Samuel Austin, was said to be the first of the travellers, although there were some who argued there had to have been others before him. Really there was no way of knowing for certain, but Samuel was the reason the travellers started working together.

Had it not been for Samuel neither group would exist. It was his choices that led them to where they were, and one of the possibilities Ben knew of was to go all the way back to then. The problem, of course, was there was no way to be certain what they'd read was the whole truth. Should the time come when Mom was written about she'd make certain no one knew about the darker side of her, keeping people from ever knowing what she was capable of.

Even journals, written by the person in question, weren't always the whole truth. As Ben flicked through the pages, trying to get a feel for the man his father never respected, it was easy to see how they'd got to the point where it seemed like they were never going to be able to find a way to work together. Constantine Austin wasn't actually meant to become the leader of their group. Instead it was meant to be his older sister, only she'd gone missing before it was possible, another sign of who he was.

Unsurprisingly there was no mention of Marie Austin. It didn't appear Constantine ever thought about her, ever doubted the fact he was the right person to be leading the travellers, even when things started to become complicated. The disconnection started before Constantine, but he knew very little about it until it started to affect his life.

'Word has been passed to me that Edward is unhappy with the situation as it is, pushing his belief we, as travellers, are responsible for the choices we make when we travel through time. He believes we need to put in place a system to keep people from going to any time they may wish to, curtailing their freedom to make what they think are the right choices.

'Having read through some off the more recent journeys certain of the travellers have made I am beginning to think he may be right. We have three new paradoxes to solve, if it's even possible, something they seem to believe is nothing more than one of the disadvantages of time travel. They don't see the hard work that's done by people like Edward to fix the damage being done by those who are foolish enough not to think.

'This is the point where Marie's disappearance becomes more of a complication. I know there are those within the travellers who believe I had something to do with what happened. Edward, fortunately, is one who knows the actual truth, and has been working tirelessly with me to free Marie from where she is most likely trapped. Every time we fail I ask myself again whether I should keep trying. Maybe it's simply not possible. Maybe she made the decision it was the best option for her.

'Before it happened we weren't close. Marie was Father's favourite, something I can understand. Of the two of us she was the most skilled, yet for all her skills she still ended up creating what seems to be an unbreakable paradox, and every time I think of her there I have regrets I try not to think about. Going back through this journal it would appear I never think of her, but I know the time will come when others read this. My own personal journal will be kept safe elsewhere, for her to read if she is ever able to return.'

Marie Austin was trapped in an unbreakable paradox Ben's father was trying to break. Standing, leaving the journal where it was, he paced from one side of the room to the other. There were no mentions of Marie in the journal he'd read, but then it wasn't one from the time when his father was with the travellers. Instead he started something new.

Could he still have been trying to break it

even after they split apart? Was that the reason for his disappearance? He'd found a way to get to Marie, but hadn't been able to find his way out. What would happen if they worked to break that paradox, to free Marie from whenever it was she was trapped, although... Ben stepped back over to the table. Paradoxes were complicated. All of them knew something about them, taught to avoid them at all costs. Yet avoiding them didn't stop them from being created sometimes.

Without stopping to think what he was doing, because if he thought he'd stop himself, Ben dialled a number. They answered seconds later. "We need to talk."

CHAPTER 3

"You know about Marie Austin?"

Mom laughed, their eyes meeting across the table. "Of course I know about Marie. Your father's obsession. I assume you've found a way to read Constantine's journal."

"I have."

"Good. I was waiting for you to make it happen." Mom tapped her fingernails on the table. "You've always been more like him, and sometimes it's hard to be around you. I know you see it as me preferring Alex. Even though this may not be something you're ready to believe I don't like either of you better, but if you're doing this then you're taking him with you."

"Why?"

"For now I have no reason to trust the travellers. Even Peter. He was always someone who fought on your father's side, and we assumed he'd come with us. When he made the decision he was going to stay, told us both he wanted to fight from the inside, we didn't believe for a moment he was going to win. Apparently he did, although we both know what the situation is. The travellers don't like being controlled, even by one of their own.

"They hate us even more, because we made the decision to walk away. It was the right choice. At the same time it was one of the most difficult choices we could have made. Edward believed we had to do it, in order to protect the travellers from themselves. He was right. Constantine believed, up until he was asked to stand down, finding Marie was going to be the answer to all our problems, the same way Edward did. Neither of them were willing to listen to alternate arguments.

"Neither of them knew Marie. I did. We were friends, for a while, before I started to understand why she was, and then... I haven't mentioned a word of this to anyone. Had I told Edward I'm certain he would never have forgiven me. You, however, might be the person who can hear me out, rather than simply assuming I was jealous of her."

"Jealous?" Ben studied Mom. "You've never been jealous of anyone."

"Edward would have said otherwise, when I got angry with him for spending all his time on her." Mom raked a hand through her hair. "We were starting to work toward his goals, and he was entirely focused on her, something I couldn't forgive. That was the point we started to draw apart, only I learnt I was pregnant. It was three years after we left the travellers.

"Like an idiot I thought Edward would start focusing more of his energy on us, because we were his family. Only he didn't I went through that pregnancy alone, before giving birth to you. My first son. His first son. I know you want to believe he was some kind of hero, but he wasn't. Your birth, it seemed, gave him more of a reason to try to get Marie out, which turned out to be because you were a reminder of how many other people his choices could affect.

"Breaking a paradox… you know it as well as I do, Ben, and if you've been spending time in the travellers library I don't doubt you've read their books too. Edward accepted it was going to be complicated, but thought it was the only way to undo so much of the damage we'd done. What he didn't understand was who Marie was.

"Honestly, I can't be certain she is still trapped in that paradox. When we worked on the calculations I changed out a couple of numbers, to make certain she would be trapped, so Constantine would be the one to take over. I couldn't be certain he was the right choice, but he was definitely going to be safer than Marie."

"Tell me about her, Mom."

"From the beginning Marie wanted to change the world." Mom shook her head. "Her vision was one where nothing bad ever happened, saving people from themselves, because she couldn't accept changing them could easily cause more problems than they'd solve. You've seen it, I'm sure, with some of the choices the travellers have made, doing damage to different parts of time.

"We worked it through, multiple times. I wanted to show her what would happen if she made that choice. How it would splinter time, in ways I didn't believe she wanted, but there was nothing I could do. She believed all the calculations were wrong. We'd made a mistake somewhere. In the end I trapped her, creating a paradox she couldn't escape easily."

"Easily?"

"Marie was one of the best travellers I'd ever met. I had no reason to think she wouldn't find her way out. As she hasn't reappeared there's a chance she's still there, or she's biding her time, waiting for the right moment to make her move. Likelihood is it's the latter. She was always patient. Spending time in the paradox before returning will also lead to her coming back young."

"She was waiting for you to get too old to stop her."

Nodding, Mom smiled. "Fortunately for me I have two sons who I know are at least willing to give me the benefit of the doubt. I know you have no reason to believe any of what I've said is the truth. This could all be lies, leading you down the wrong path, so Alex can trap you in a paradox. I'm not going to lie about having had the thought a couple of times.

"Our relationship might never be what it could have been, but I trust you to make the right choices. You went to the travellers to learn more, to hear both sides of the story, which might actually be my saving grace. Talk to Peter. Ask him about Marie. Out of all of us he was the most neutral when it came to her, and he's probably the one traveller still there who actually knew her."

"What happened to Constantine after he was asked to stand down?"

"I don't know. Peter might. If he is still around there's no doubt in my mind he's going to be trying to find a way to save her, because he couldn't think of anything else. He believed she was the one who should be in his place. What he didn't see was the dark path she was going to lead the travellers down."

"Splintered time is repairable."

"When the splinters are small, yes. It's something we've all learnt how to do, to protect everyone around us from the travellers, but Marie would have broken it into hundreds of pieces. Saving us from that fate was my priority."

Alex walked into the coffee shop, moving over to join them. Like normal he took the spot next to Ben, even though it meant shoving Ben along the bench, grinning as he did. Ben shook his head. Alex was never going to change. He was still a big child at times, especially when it came to screwing around with his brother.

"Ben knows?"

"He does." Mom looked between the two of them. "He's going to be staying with the travellers for now. Peter may even be willing to help you, if you share with him the story I told you, because he will remember me. I'm not going to say we had the nicest of goodbyes. I was angry with him, and he knew it. At the same time he knows me well enough to understand why I made the choice I did.

"Out of all of us he was likely the one who understood the most about splintered time. During out early years as travellers he was the reason I learnt as much as I did, due to his belief no traveller should journey through time without understanding the damage they might do. He's the reason I knew Marie's plans were going to destroy time, in order for her to create one where nothing bad ever happen."

"Peter can be trusted?"

"For now I'm going to assume yes, as he didn't kick Ben out the moment he learnt the truth about who he was. Edward's son back in the travellers..." Mom smiled. "When it comes to both Marie and Constantine there was the Austin graveyard. If there's anything personal of theirs still around it's likely to be there. Get lucky, and you might even see Constantine."

"Who's probably dead." Alex shook his head. "We can't be certain of anyone yet, Ben, and I don't want anything to happen to you while you're with the travellers, which means making the logical choices. Limited trust."

"Yes, I know." Ben's voice was flat. "I'm not an idiot."

"That's news to me, considering the fact you walked into the travellers' without stopping to think about backup."

"I have backup."

"Not me."

"Choosing you as my backup seemed tantamount to suicide. I knew better than to do something that stupid."

"However angry I've been with you there was never a time I was going to trap you in a paradox."

"Okay, that's enough." Mom raised an eyebrow. "This animosity between you is my fault, and I want both of you to think before you start bickering. Neither of you have a reason to trust each other, any more than Ben has a reason to trust me, which is something you have to keep in mind, Alex. Ben is giving us a chance we technically don't deserve. I know what it's like to look at things from the outside, the way he did."

"Mom..."

"No, Alex, you have no reason to see this decision as a slight. It was Ben making the logical choices, to protect himself. Especially as he sees the arguments going on within our group, with me never stepping forward to choose a side, seeing that as my way of being on both of them at the same time."

"You aren't?"

Smiling at Ben, reaching a hand across the table to touch his, Mom shook her head. "My neutrality, like Peter's, is my power, and the moment I choose one of them over the other I tip the balance. For now I'm able to force them to listen, even when they don't want to. I can stop either side from doing anything truly stupid.

"In time you'll come to see it, but you won't be alone. My hope is that you'll be able to build a relationship with Alex, so the two of you have each other's backs, because that's what you need. Especially now. Marie is our greatest danger right now. Then we deal with the travellers who no longer understand what it means to be what they are. Maybe we'll even be able to find a way to work together."

"Probably not, though."

"Alex, that's not necessary."

"He's right." Ben shrugged. "We might never be able to find a way to work together, but we do need to deal with Marie, if you are telling me the truth."

"Like I said Peter will be able to tell you from his point of view what she was like. I trust him to be honest with you, even if he doesn't necessarily see her as a problem. Like Edward he was close to Constantine. The three of them were friends from the time they started learning what it meant to be a traveller. It was the first time Constantine had anyone in his life who wasn't Marie.

"For the first time in his life he wasn't being compared to her, and being found lacking. They wanted to be his friend for who he was. Everything that happened between them later on changed their relationship, but they always cared about each other, which was why Edward kept trying to save Marie. It was his way of apologising for his decision to leave, because they couldn't find a way to compromise. Only it also meant he ended up losing me, along with his chance to be a father to the two of you, something I hope he regrets."

CHAPTER 4

Running his tongue over his bottom lip, seeming a little disconcerted, Peter studied Ben. "Marie is a complicated subject, for a lot of reasons."

"Something I understand, but I need to know more about her."

"Yes, I'm sure you do, if you're read what I think you have." Peter breathed in deeply. "There are certain parts of our history most know nothing about, with Marie being one of them, along with Edward's lifelong search for her, until the time came when he was also lost. I wouldn't be all that surprised if Katherine did have something to do with Edward's disappearance if he was still trying to break Marie out of the paradox she was trapped in."

"Why do you say that?"

"This means going back a long way. We all

have personal connections that become important. Edward was one of my closest friends, along with Constantine. Their conversations were something I was never a part of, in part because my decision was made long before Edward's - I was staying, to try to change the travellers from within. He didn't believe it was possible, which was why he ended up splintering off.

"I doubt he mentioned me in his journal. Our friendship mostly faded away after he left, but I never stopped thinking about him. Learning he was lost... I wish I could say it was a surprise. In the end it was an inevitability. He always put the search for Marie first, over everything else, believing he was the only one who'd be able to break her out. The reality was very different.

"Edward was skilled. Katherine was better. She was probably one of our best, and her loss hit Constantine even harder than Edward's, especially as he viewed her as his last link to Marie. The two of them were close, before Marie disappeared. Out of everyone Katherine was the one who knew Marie the best, so when she tried to talk to the two of them out of what they were doing I listened, before trying to work out what they might be the case."

Ben nodded. Confirmation Mom was telling the truth about her friendship with Marie was a step in the right direction, even if that didn't necessarily mean she was telling the truth about everything. "What did you learn?"

"Marie was making dangerous choices." Peter shook his head. "Katherine didn't want to talk too much about it, but she left Marie's work journal in the right place for me to find it. We all knew Samuel planned on Marie taking over. She was far more skilled that Constantine, which was part of the reason she ended up friends with Katherine, because they were able to push each other.

"However that did lead Marie down a dangerous path. She wanted to show what a traveller was truly capable of, believing it was possible for us to undo some of the worst atrocities the world has ever seen, which... having gone through everything she'd done it's far more likely she would have completely splintered time, possibly making it so there was no chance of anyone being able to put it back together. Her disappearance saved us all from that."

"What do you think happened?"

For the first time Peter smiled. "A couple of numbers in the wrong place was all it took, Ben. Katherine knew exactly what to do. If it wasn't for the fact I knew what Marie was planning I never would have suspected Katherine, but seeing that... Katherine did what she had to do." The smile faded. "It hurt her, in ways she was never able to let out. Had she been able to truly trust Edward it might have made things easier.

"Instead she had this huge secret, one she couldn't tell anyone, which easily led her down the same path Edward walked. Out of the two of them she was the one who believed more deeply in the dangers of freely travelling through time, and she was willing to do whatever she had to in order to protect time from, as she put it, all the idiots."

"Marie included."

"Letting her best friend off wasn't something Katherine would ever have chosen to do, however much she might have wanted to. She had to protect those who couldn't protect themselves." Peter shrugged. "I understand her choices, even if I don't agree with all of them."

"How likely is it you think Marie would be able to find her way out of the paradox without any help?"

"Greater than Edward's chance of being able to get her out." Peter's eyes met with Ben's. "Now I think you need to tell me why we're having this conversation."

"Mom thinks Marie's been biding her time, and I should be preparing to deal with her sooner rather than later."

"Entirely possible. I wouldn't ever ignore something Katherine was telling me, especially now. She has a reason for thinking it might be soon." Peter stood, stepping over to a calendar hanging on a wall. "Of course. We're a couple of weeks away from a new leader being chosen. Marie could easily take over if she was around, because she's an Austin."

"We need to find Constantine."

"Do we? He's not going to help us stop Marie." Peter stilled, looking back at Ben. "He had to have known. There's no chance he couldn't have done." He raked a hand through his hair. "Marie's journal's would have been left behind, and she would have written this down, yet he was still determined to get her out."

"He could have believed he'd be able to talk her out of it."

"Yes, he could, but Constantine wasn't someone Marie ever respected. If it had been down to her he never would have taken her place. Fortunately it was, it was Samuel's choice, and he gave Constantine a chance. Some people believe Constantine failed Samuel, when it truth he did everything Samuel could have wanted.

"Samuel was a deep believer in control. It's why he gathered the travellers together in the first place. Seeing what's happened to us now would have him turning in his grave, not able to understand how it is we've got to this point. Constantine's choice to stand down had an effect. There were no other Austins, because he was so focused on getting Marie back, and sometimes I think he was never going to be able to accept Katherine was with Edward."

For a few seconds Ben was actually lost for words. “Mom and Constantine?”

“On Constantine’s side, never Katherine’s. She chose Edward, and Constantine pined from a distance, wishing things were different. Had he been more capable of opening up to her he might have changed the future. Instead he kept his feelings to himself. If he ever knew the truth about what happened to Marie he never would have forgiven Katherine for her part in it.”

Nodding, Ben raked a hand through his hair. “I still want to talk to Constantine.”

“Most of the time he’s hard to find, but at this time of the year he’d be at the Austin graveyard, somewhere I’m certain Katherine would have told you to go.”

“Would she have known?”

“Katherine, I think, knows everything, one way or another.” Peter laughed, standing. “Let me take you there. Getting this conversation out of the way sooner rather than later would definitely be the most logical thing to do. Then we can start looking into whether it’s possible

Marie is somewhere around."

Ben followed Peter out of the library, through a different door that led them outside, along a path that looked well worn. When Ben glanced back at the door it was barely noticeable. It was likely only someone who knew it was there would be able to see it. As they continued walking he realised he was somewhere he'd never been before, going down what seemed to be some kind of hill, that led them toward a small graveyard.

Inside a man sat on a bench. As they got closer to him he looked at them, his eyes tired. There was no doubt it was Constantine. Ben saw pictures of him before. Standing, Constantine moved to the gate, looked at the two of them in turn. His eyes finished on Ben, studying him.

"You look like her." Constantine breathed in deeply. "Sometimes she comes by about this time, but I'm not surprised you're here now. Like normal she's probably right. I didn't truly accept what she said until Edward went missing, and then... it's hard to accept the sister you've always looked up to, wanted to be like, is actually a danger to everything you've been working toward. That was the time I made the decision to stand down, although the travellers made it easy for me. They believe they asked me to.

"Marie... she was complicated. You're read my journal, so you'll know we were never close. To her I was little more than nothing. Katherine, on the other hand, was exactly the kind of person Marie wanted in her life, only she didn't realise Katherine would never agree with the choices she was making.

"Katherine was the daughter of two parents who had no link to any of us. It's not unusual for that to happen, which is why Father wanted to find a way to get us together, so we could learn what it meant to be a traveller. Yet she was the one who was more like Father than anyone I've met. If I wasn't certain it was impossible I'd have said she was Samuel Austin's actual daughter.

"Then there was Marie, who wanted to push the boundaries, to see what travellers were truly capable of, going against everything Father taught her. I think that's why it was so hard for me to accept it was possible that was the path she walked. How she made the decision to give up everything we were, even after she was shown the damage she might do. Splintering time, apparently, didn't bother her at all, something Father taught us both to avoid at all costs."

"My dad's the reason you changed your mind?"

Constantine nodded. "Edward was a good friend, even when the two of us couldn't find a way to work together. I… being the leader of the travellers was far more complicated than I could ever have anticipated. Father explained it to me, but hearing it was nothing compared to living it, and I was trying to find a middle path I don't think existed. At the time I wanted to believe it could.

"Only Katherine really accepted it early on. She tried to get me to see, and I wasn't willing to hear what she was saying. In part it was anger, and jealousy, for her choosing Edward over me, especially as he was choosing Marie over her. Well, to her, it was probably more him choosing me over her."

"Why now?"

"We're coming up to an interesting time in Austin history, something Katherine knows well. Her friendship with Marie taught her a lot about the family. This is the 60th anniversary of Father starting to gather the travellers together. For Marie it would make the most sense to take it back now, to give her control over everything. After everything she's been through to get here I believe she thinks she deserves it."

"Tell me how to deal with this."

"From the beginning." Constantine looked at one of the gravestones. "The Austins have been travellers for a long time, since long this existed, but building up this... Father believed it was the right decision to make. Maybe he was right, and I was the one who failed him, because I couldn't hold it together. He would have done."

"You have no way of knowing that." Ben shrugged when Constantine glanced at him. "We look up to those around us, seeing them in a certain way, and yet something they aren't who we believe they were."

CHAPTER 5

Sitting at a table with the group he was with wouldn't have seemed possible when Ben first walked in to the traveller's headquarters. Maybe it was always an inevitability, one he simply had no way of knowing about, because the history he knew wasn't the whole truth. Instead there was much more to it.

Constantine was the least comfortable, eyes on the table. He hadn't looked at Mom since they'd entered the room, which was almost amusing, although she was studying him, teeth buried in her bottom lip. Finally she reached out a hand, putting it on his. Neither of them said anything for a few seconds. They probably knew they were being watched.

"I forgave you a long time ago, Constantine." He lifted his eyes. "I understand the choices you made better than I ever did before." Mom looked at Ben, and then at Alex. "I've made my own mistakes, ones I can't undo, something I can only hope I won't live to regret."

"With them both being here I don't think you will." Constantine smiled. "Do you ever wonder..."

"Yes, but I made my choice." She shrugged. "I loved Edward, as hard as it could be at times. The feelings I once had for him no longer exist, not in the way they did before, yet I'm also going to remember what came from the moments we had together."

"Marie..."

"Even if you hadn't spent half your life trying to save her she'd still be a problem. That's who she is. Marie and I were friends, but we never saw things the same way, even though she spent a lot of time trying to convince me she was right. Experimenting was our right. Listening to her was painful, knowing what Samuel taught you, and what I'd learnt myself, seeing the damage she could have done even if she didn't end up splintering time."

"How likely was that?" Alex looked between Mom and Constantine, the mix of emotions in his eyes obvious to Ben. "Splintering time with her experiments?"

Both Constantine and Mom were silent for a time, before Constantine breathed in deeply. "Almost certain." He turned to Alex. "Edward was more willing to accept that than I was, but he still believed we'd be able to talk Marie out of doing what she planned. Katherine argued otherwise. She told us we were making a stupid mistake. There was no chance of us ever being able to make Marie see sense, because she hadn't, and Marie was far more open to hearing her out than she was the two of us.

"There was also a part of me, one I couldn't ignore however hard I tried, telling me I couldn't give up on her. She was meant to be the person in control of the travellers. Had she been there she'd have been able to deal with everything far better than I could, maybe seeing the disconnection between the two sides as a reason for her to keep following in Father's footsteps."

"Wishful thinking." Mom sighed. "Edward was the same. You both believed in her in a way I couldn't because I knew who she was. She was open with me, in a way she never chose to be with anyone else. Not even Samuel. Of course she knew if she ever mentioned what she thought to him he'd do everything he could to stop her. In the end I was the one who had to, trapping her in a paradox."

"Until now." Ben raked a hand. "Putting off the problem to another day."

"Possibly. I can't be certain..."

A rumble went through the shop. Ben grabbed Alex's arm, dragging him away from the table, right before a jagged hole appeared where they'd been seconds before. Mom, Peter, and Constantine were on the other side. From the look they were giving the hole it was obvious the possibility had turned into a certainty, especially when it disappeared moments later, all of them back where they'd been before.

Blinking, Alex shook his head. "What just happened?"

"A splinter." Ben looked at his brother. "Relatively small one, considering how bad they can be, but still enough of a sign."

"How do we stop this?"

It was Constantine who shook his head. "Not easily." He pulled a journal out of his pocket. "This is Father's journal, from fifty years ago. I've been going through it for days now, thinking about how Marie could be stopped, and I'm certain it's not by going back. Doing that will cause more problems than it solves." He put it on the table. "By doing what he did Father changed the lives of multiple people. Every one of those people is going to be connected to others.

"None of us would be where we were if it wasn't for Father. I wouldn't be alive, and neither would Marie. That could stop all of this. At the same time we know there is a chance of making things worse than they were before. Marie might still be born, without Katherine there to stop her, leading us to this position far sooner. Only it wouldn't be us. It would likely be an entirely different group."

"Hence the arbitrator." Katherine looked at Peter. "The one who tracks all the possibilities when someone goes through time." She raked a hand through her hair. "Have you got anything to say?"

"Nothing useful." He almost smiled. "I wish I could say otherwise, but Constantine's right. You not being here would cause the biggest changes. There is also the possibility of Marie trapping you before you could trap her, something we know could happen." He pulled a notebook toward him. "When it comes to a journey, and a decision as big as this one, I go through all the likely timelines that could have existed due to the choices that were made.

"With this we have our current time line. Katherine is alive, Marie was sent through time to stop her, and we all got to this point. Then there are the other options. Katherine never lived. There could be someone else in her place, who is exactly the same as her, which is unlikely. However that's another good possibility.

"Then we have the person isn't like Katherine, and works with Marie. Bad. We have them being like Katherine, only they were trapped the way Marie was, while Marie went through with her plan. Bad. We have them being someone who fought with Marie, and never had a chance to stop her. Bad.

"So, with one possibility, we have three bad to one good, and this is just with me going for the most basic options. If I was doing this elsewhere, with more time, I'd have more possibilities. Often you find that the bad over weighs the good. That's why I don't often approve of journeys that will change anything, especially if it's tightly entwined in our future."

"Marie, on the other hand, doesn't care the same way we do about our current future. She never did." Mom looked at the notebook. "She wanted to understand time in a way no one else had done. How different futures would affect people. I explained to her the damage she'd do, in multiple different ways, but she wasn't willing to listen, because she'd already made her decision."

"Can we stop Marie from being born?"

"No." Peter shook his head, looked at Alex. "We have no control over the future we create. One of the reason travellers get trapped in paradoxes is because they create a future they no longer exist in, so they can't get back to it in order to undo the damage, and therefore find themselves lost within something time makes to protect itself.

"What Marie's doing is likely to create paradoxes, but she's got the skills to break through them, to travel to futures where she no longer exists, which is one way to splinter time. There are others. Everything we're talking about now has other possibilities, and there's no way of knowing how bad things could be.

"I can tell you now that we're in a splinter where we do still exist, and Marie does too. We will continue to exist in this splinter now we're here, so we don't need to worry about how many there are, just that there already is one. Dealing with that splinter, if it had nothing to do with Marie, would be simple enough, only we all believe she made it."

"How did she make it?"

"That's the more complicated question." Peter ran his tongue over his bottom lip. "There are books in the library that track time changes. Samuel crafted one of them, believing it would be important in the future, and there are others, copied from what he did. They exist to make it simpler for us to deal with splinters.

"Only problem is we don't know who we are. Being here... it complicates matters a little." He looked at Constantine. "There's one person here who can walk into library without anyone stopping him. Not even Marie would be able to, if she has already found a way to take her 'rightful' place as leader of the travellers."

Constantine nodded. "The rest of you could wait in the graveyard. I'll grab a book, and head out as quickly as I can. Unfortunately, depending on our future, that might take longer than any of us want."

Ben shrugged. "Now we have a splinter time doesn't matter exactly the same way it did before. Yes, it might take longer, but putting the splinters back together is going to affect this future, and all the others she might already have created. We all know it."

"You're right, and this future isn't going to change. At least we don't believe it will. Father believed there were those who would be able to travel from one splinter to the next, which is something we also need to keep in mind. Marie might have been able to learn how to do it. If she has..."

"She can't." Constantine looked at Mom. "It's something she attempted in the past, and couldn't make it work. Another of the many things we argued about." She shrugged. "Edward might be able to, but we have no way of knowing if he is involved with this now, although it is a possibility."

"Edward working with Marie?"

"If it got him free of a paradox I can see him doing it. He wouldn't have been able to get out alone." Mom shrugged. "The problem was I had no idea when he was, so I couldn't go to free him, and there was no logical reason for me to attempt to check Marie's. Unfortunately he was never as good at keeping records as I wanted him to be. We talked about it a lot. I told him if he got stuck somewhen I needed to know when it was in order to get him out.

"Going through what he'd done didn't tell me anything useful, but I did check the times I could, doing my best to find him. Of course I was also a mother to two young boys at that point, which made it much harder. Talking to the others wasn't something I was willing to do, in case he did ever get back, because then he'd be able to take over once more."

Sharing a look with Alex, Ben turned back to Mom. "You were going to simply let Dad have control, after being away for over twenty years, as though people would accept him."

"Most already knew him, Ben. He was the one they started the journey with. I have no reason to think they wouldn't have jumped at the chance to have him back."

"Your future must be different to this one." The voice came from the door. "Pleased to hear you're still the person you once were, Katie, even with me being such an idiot. I never deserved you."

CHAPTER 6

Constantine was in the library, hopefully grabbing the book they needed. Ben looked at his father, trying to wrap his head around everything, because Edward didn't look much older than he was. Alex seemed to be doing the same thing, head tilted to one side. Maybe he was doing a deep search for wrinkles.

"I know this is weird." Edward shrugged. "I never thought I'd get back, so I'd given up on this being a possibility, until Marie dragged me out, creating a splinter as she did. Unsurprisingly she was a decade younger than me, but she still seemed to think there was something we could build between the two of us." He shook his head. "Letting her think that seemed to be the best way of getting back here."

Slowly, Mom nodded. "How did you get here?"

"Unlike Marie I can jump splinters. Getting away from her was the easy part. Finding you was far more complicated, especially when I was also looking for the us who had two sons, and Marie's choices are creating more than one splinter at a time."

"How many times did we get together?"

"Three, maybe, with two of those not ending with children. Mostly because you gave up waiting for me to change. Learning how much pain I'd put you through... I'm sorry, Katie. Apologising now isn't going to change anything, and I can understand if you aren't willing to forgive me, but I still want you to know I regret the choices I made."

"You were forgiven a long time ago, Edward. I don't hold on to grudges. The choices you made did change my life, and I was angry for a while, especially when I was traipsing through time trying to free you. Working through it took time, but I'm not still angry with you, any more than I'm still angry with Constantine for getting you into it in the first place."

"Can we..."

"No, because I'm twenty years old than you."

"Do you really think I care?"

"Probably not, but that doesn't mean I haven't had twenty years to get over you." Mom looked at Ben and Alex. "Twenty years to raise our sons alone, because you put what Constantine wanted over our needs. I'm more angry on their behalf than I am my own. You have up on having a relationship with them."

"Making the choices I believed were right." Edward sighed. "I thought I'd have time, Katie, to make up for everything, and now... well, obviously I don't. I made bad decisions, but there's no way of changing them now."

"Unfortunately not." Mom stood, stepping over to the gate, and Ben turned to see Constantine walking toward them. "How bad is it?"

"We're coming up on over a hundred splinters created, with Marie still going. It's not going to be long before we start seeing problems arise from her choices."

Ben stood. "What has she changed?"

"Small things for now, mostly linked to the travellers, probably to see what she can do. Then she'll start going further back, changing bigger events. Doing that will make the splinters harder to put back together. It why Father never let anyone change historical events, because the moment that happens we start seeing problems everywhere, even in the splinters where it still happened. Fortunately we are still when we were, with the only major difference between Edward.

"However that change is because you made the decision to find us, Edward, when Marie found herself in whatever splinter it is she's on now, which gives us a chance to deal with her first. She's trapped. If she wasn't it would make this far more complicated, because we wouldn't know which one she might be on, but this makes it possible."

"As long as someone else can splinter travel." Katherine looked between Ben and Alex. "It's going to be one of the two of you, if it's possible, because Edward can do it. This is something that's normally passed down from one generation to the next. Unfortunately I can't, so it's going to be up to him to teach you how."

Edward stepped forward. "Until today I didn't know I could splinter travel. All I knew was I wanted to be away from Marie, and it happened. From there it was more complicated. Working out how to use this skill purposefully took more effort. The last thing we want is to reach Marie straight away, especially as it's likely she's bouncing through time. Travelling both in time and from one splinter to the next isn't going to be a simple process.

"Close your eyes. Think of another graveyard, with no one there. We want to be entirely alone, rather than jumping somewhere someone might be able to see us, but there is a chance neither of you will be able to make the journey. There's no way of knowing until you try."

Glancing at Alex, who nodded, Ben closed his eyes, thinking of the graveyard. Rather than Mom, Peter and Constantine being there no one was. Breathing in deeply, Ben opened his eyes once more. Well, he was in an empty graveyard, but he wasn't with Alex or Edward, which probably wasn't the most useful thing.

Instead of closing his eyes Ben saw the graveyard he'd just left in front of him, and then there was a rush of everything all around him. Then he was back where he had been. Neither Edward or Alex were there either, which at least was a sign they were all able to travel between splinters.

Mom stepped forward, looking Ben over. "Are you alright?"

"I'm fine. It's not too different to travelling through time." He smiled. "You don't need to worry about me."

"Telling me that isn't going to change anything. However old you get I'll worry about you, because you're my son." She wrapped her arms around him, pulling him into a tight hug, something she hadn't given him for a long time. "I don't show you often enough how much you mean to me, Ben."

"Mom..."

"Yeah, I know. You're too old for all of this." She laughed, kissing the top of his head. "That isn't going to stop me either." Moving back their eyes met for a moment. "We'll find a way through this, one way or another."

Alex was the next one to appear, shaking his head as he did. “We all went to different graveyards, because more than one of them is empty, so that worked well for a test.” He looked at Ben. “You back already, or did you never go?”

“Back already. There was no point in hanging around.” Ben shrugged. “I came straight back.”

“So did I.”

“Different splinters will take longer to travel between.” Mom smiled. “At least that’s what I was told, by another splinter jumper, a very long time ago.”

“Who else can jump?”

The smile faded, Mom running her tongue over her bottom lip. “Unfortunately he’s no longer around. Trapped in a paradox, at least according to official records, even though he was one of very few jumpers we had. Someone wasn’t careful enough with the calculations they gave him, and we haven’t found a way to get him out safely yet.”

Finally, Edward appeared. “How did it go?”

"Both of them can jump." Mom looked at Edward. "What happens next?"

"We start the complicated part, which means jumping to the splinter I was at. Normally jumpers travel alone, so I have idea how this is going to work, but I want to do this the safest way possible. Getting here took me multiple jumps, but I might be able to get back to the other splinter more easily because I already know it. Only that means testing out a theory."

Sharing a look with Alex, who seemed as uncertain as Ben, they both turned to Edward. "What theory?"

"If we're touching we should be able to go to the same splinter." Edward held his arms out. "Are you prepared to try doing something different in order to save time as we know it?"

"We aren't superheroes, Edward." Alex rolled his eyes. "Let's go."

Grabbing Edward's arm with his hand Alex looked at Ben. "What happens if this goes wrong?"

"That's a question I can't answer." Edward

studied Ben. "I'm assuming nothing, but I could be very wrong. Unfortunately we don't have a lot of options right now. If we're going to stop Marie then we need to make the hard choices, which I'm assuming you were already willing to do, considering the situation you're in now."

Nodding, Ben took Edward's other arm, before Edward grabbed both their wrists. Alex reached out for Ben's other arm, creating a three way hold. Ben closed his eyes, knowing Edward had to be the one to envision the splinter they were going to, keeping a tight hold on his brother and the man he'd been planning on trying to save. Only now it was so much more than he thought it would be.

After a minute, maybe a little longer, Ben opened his eyes. They were definitely in a different splinter, but there was no way of knowing if it was the right one. Until a rumble caused him to jump to one side, watching as both Edward and Alex went to the other side, safe from falling into the hole. Normally it would have filled in straight away. Instead it stayed right where it was. Blinking, Ben stared at it.

"Okay, so this is what happens when someone creates too many splinters."

Edward nodded. "You begin to cause damage to the world. This won't be fixed until we find a way to stop Marie, which means going to talk to the travellers. I don't know much about this splinter. I have no idea what kind of situation we're going to find ourselves in when we do, but it's likely they already know something bad is happening, so we should start there."

"We're trusting the travellers."

"You were already trusting them, Alex, when you chose to start working with Constantine and Peter. This is no different."

"It's different." Constantine looked at them from the gate. "First you have to convince me to trust you." He ran his tongue over his bottom lip. "Considering the choices you made before, Edward, I have no reason to think that will be an easy thing."

"Maybe not for him, but we aren't Edward."

"No, you aren't." Constantine studied Ben. "You look a lot like Katherine." He ran his tongue over his bottom lip. "You look a lot like her son."

"Travelling splinters doesn't cause paradoxes?" Ben looked between Constantine and Edward. "How is that even possible?"

"Ben, you know as well as I do time travel is more complicated than it appears, and yes, you do also exist here, but Alex doesn't. At least not as my son." Edward raked a hand through his hair. "I know the choices I made can't be undone, Constantine. I was an idiot. Unfortunately there's nothing I can do about it now."

"Can we go back to the beginning?" Alex raked a hand through his hair. "I don't think I understand how it is splinters actually work, because the two of you have a whole history, which is not the history we know, and I'm really confused right now."

"Understandable, if you're jumping splinters." Constantine gestured for them to follow him. "I will give you both the benefit of the doubt, because it's obvious you aren't here to harm me, and you came with Edward, so there has to be a reason for that. Apparently because he thought he could apologise for being who he is."

"Fair." Edward sighed. "I did think that was going to be a possibility."

"Katie doesn't hold grudges, Edward, however angry she was with you when it first happened. She's made her peace with things. Fortunately for her she had people around her who were able to help."

CHAPTER 7

Katherine looked at Ben, and then looked at her son. There were no differences between the two of them physically. Yet there was no doubt they were two very different people. Ben raked a hand through his hair, staying closer to Alex, because it made everything a little easier. Neither of them knew enough about what they were doing, but it wasn't as though they'd had any reason to learn about actually being able to travel between splinters.

"How about you start at the beginning, Ben?" Katherine gestured for them to sit down. "Splinter travel is complicated at the best of times, and this is definitely not the best to times."

"There's a nice new hole in the graveyard." Constantine sighed. "From what I heard you saying you at least know why it is we're dealing with them."

"Marie." Ben ran his tongue over his bottom lip, looking at other Ben, who nodded. "This is the first splinter she created, so I think it would help to know what is different here, compared to where we came from. In ours there are two groups, the travellers, who are dealing with a lot of problems right now, and us, dealing with different problems.

"I made the decision to see what I could learn from the travellers, which led to me hearing about Marie for the first time. Until more recently she wasn't a problem, but Constantine believes this is a time when she would be most likely to try to change the world, due to certain anniversaries that are coming up.

"We believe this Marie is the one who was trapped in the paradox Mom created there, trapping Marie so she wouldn't be able to cause any more damage, even though Mom didn't believe it was possible it would keep Marie trapped. She, I think, was always planning on me doing what I did, as it meant I learnt a truth she'd never told me."

"Sounds like me." Katherine shook her head. "Marie, fortunately, wasn't too much of a problem here, because I was able to talk her out of walking that path. Instead that was Edward." She looked at him. "Something I didn't learn until after I got pregnant, and I made a very difficult choice, because I couldn't let him do what he planned."

"Planned with Marie."

"Yes, you said that, which is why Father made the decision I was going to take over instead, even though she tried to talk him out of it. It was something Marie never forgave us for, especially when we put her under strict controls to stop her from being able to travel through time freely. That was the point the split happened." Constantine sighed. "There was a large group who wanted more freedoms, so they chose to take it. We've been working to keep them from doing too much damage."

Alex and Ben looked at each other. "The opposite of how things were before." Ben turned to Edward. "You knew everything was different where we were, because you went to find some allies."

"Until I realised who I was there. Seeing myself in Constantine's position was interesting, to say the least, as I tried to work through what it all meant. Now..." Edward shook his head. "Obviously it doesn't matter. This is how things are. I can't change what was in your splinter, and I wouldn't want to, because I know this one won't exist soon enough. The person I am now isn't real."

"The complication of splinters." Katherine shrugged, looking at Ben. "I'm assuming I'm the current leader of the group working to stop the travellers from making stupid choices." Ben nodded. "Well, I'm not in a very different position now, although it's us who's stopping the other group, due to the different split. I suppose it makes sense. I was always going to walk that path, however I had to do it.

"I have two sons." She looked at Constantine. "Both of them with Edward, rather than having future children with Constantine, after getting close when he was helping me with Ben. He took you on as his own, which was always what I expected. What happened to your mom?"

"She raised us alone, when Edward got trapped trying to save Marie. Constantine wasn't willing to accept his sister was going against their father the way she seemed to be, and it led him into making bad decisions, before giving up entirely on being the travellers leader after Edward disappeared. That was when Mom took over the group, doing what she had to in order to keep the travellers from causing damage that couldn't be undone."

"Does any of this actually matter?" Other Ben stood, moving away from the table. "We know we aren't real. Who they are doesn't change that. What we need to be doing is focusing on a way to stop Marie, so we can start putting the splinters back together."

"Only it's more complicated than that." Constantine moved to where Other Ben was, putting his hands on Other Ben's shoulders, Ben trying to work his way through what it meant for the two of them to have the relationship they did. "How they go back together matters. We could end up being the surviving splinter, no matter what Edward said before. He's not dealing well with all of this."

"No, I'm not, because the woman I loved trapped me in a paradox, and when I finally got back to her, believing we might be able to start again, she'd already moved on."

"It's been twenty years, Edward." Katherine raised an eyebrow. "We were never going to be able to make things work, however much you wanted to, and I'm sorry I chose to find myself some happiness. Did you really think I was going to wait for you, when you were willing to destroy time."

"To make it better."

"You see what it would have done." She gestured. "The damage Marie is doing to us, jumping through time, making things better." She shook her head. "You were an idiot to think she was right. I don't regret the choices I made back then, however much you might thing I should, because I know they were the right ones."

"Katie..."

"Don't even think about trying to convince me otherwise. You could have done this without creating hundreds of splinters." She looked over at Ben. "Please don't think what Constantine said means we're going to try to take over. It's more a matter of how the splinters come back together, because this many… it's not going to be easy for us to undo what Marie's done. Something I'm sure you're beginning to understand."

Ben nodded. "Splinters were something we were taught about in order to put time back together, not to travel between." He ran his tongue over his bottom lip. "This is entirely new."

"Understandable." Katherine looked back at Other Ben. "We did the same thing. I'm sorry, love. I know this is hard. Splinters are really complicated. I can't imagine how you're feeling right now, but we will find a way through this."

"I know." Other Ben looked at Ben. "We're probably very similar."

"Honestly, I doubt that." Ben glanced at Alex, who grinned, shaking his head. "My relationship with Mom is very different to the one you have with yours. I can see that from here. You also had Constantine. We didn't have any father in our lives, which was, in part, a choice. Neither of us wanted someone to try to take Edward's place, believing for a long time he was going to find a way back to us, until the time came when we couldn't hold on to that any longer. It was obvious it wasn't going to happen."

"We didn't deal with that realisation in the same way. Ben focused on trying to find a way to save Edward, which led him to the travellers, and instead learnt about Marie, who happened to be the reason Edward was trapped. I gave up on him entirely, focusing on living my life. Course he thought I was the favourite, being taught everything I needed to know by Mom in order to be the next leader, when we were both watching him instead."

"Alex..."

"Mom wanted me to support you, no matter what. She never wanted you to be alone, dealing with problems you couldn't solve, so I was doing what I could, even when you didn't know it. Even when you were sneaking around, making your plans. I kept your secret for you, because I trusted you to be making the right choices, at a time when I knew how you saw me. I don't care about that. All I wanted was for us to be okay in the end."

"No matter what happens next, Alex, we are okay." Ben looked at him. "I'm sorry. I was so focused on me I didn't see you, and I apologise."

"It's done." Alex turned to Katherine. "Your son's right when he said we need to focus on Marie. Getting that dealt with first will stop more splinters from being created, and then we can start putting them back together. Only we know she's going to be bouncing through time, making her changes, because she believes, mistakenly, she can create a world where nothing bad happened."

"Or at least that was what she believed. We have no idea what she might be doing now, after twenty years of being trapped in a paradox. All we know is she's definitely behind this."

Nodding, Constantine went to grab one of the books. "We can see what she's already changed here, the number of splinters growing faster than I thought it was possible for one person to change time, but then I don't know how much she cares about how she's changing it." He raked a hand through his hair. "It looks like she's not really making any logical choices right now."

"She's not. Marie had a list of times we'd go

through, based on what she thought needed to be changed the most. I know she believed I'd be able to help her travel through the splinters, until she found the right one, which is why she made the decision to drag me out."

"Instead you jumped away from her, trying to find a splinter when Mom hadn't found someone else, believing it meant she hadn't moved on." Ben shook his head. "You wanted to try to build a relationship with us, in the hope you'd be able to find a place there. Maybe you hoped your other self wasn't trapped in a paradox."

"Ben..."

"You weren't helping us out of the goodness of your heart. You wanted to try to take Edward's place, as though that was possible." Ben stood. "We need that list, Edward, so we can find her, and put a stop to this."

"How, exactly, are you planning on doing that, Ben? A paradox? It didn't work last time."

Laughing, Ben shook his head. "I'm not an idiot. There are more permanent options, and I have no reason not to use it. Marie means nothing to me." He shrugged. "I'm going to do what needs to be done. That was always the plan. I didn't think for a moment that meant having to kill someone, but time is more important than one person. I'm not going to let her do this."

Everyone, apart from Alex, stared at him, like they couldn't quite accept what he'd just said. "I hate it, but he's right." Katherine sighed. "The only way we stop this from happening again is a permanent solution. Marie will try again if we don't. At least that Marie will." She ran her tongue over her bottom lip. "We need to find a way to put the splinters back together to make certain no Marie can ever do this again."

"Doesn't mean this will be over." Edward's voice drew their attention to him. "There will always be someone who wants to make this happen."

"That's why we exist, Edward." Ben stepped over to him. "We stop people like you, who want to destroy time the way Marie has. However complicated it might be we'll do our best to make certain this will never happen again."

CHAPTER 8

Ben and Alex stood with Other Ben. "Are you certain?"

"This is my splinter she's destroying. I'm willing to do this, to help you. Our history is slightly different to yours. Getting lost would be possible, so it's best we do this together, to keep you both safe." Other Ben held out both hands. "Get either side of me. I have no reason to think this will be easy. Marie will try to slip us, which means we need to keep a connection to her when we do find her. Try to grab some hair, or a piece of clothing, as long as it's hers it should work."

Nodding, Ben stood one side of his other self, their eyes meeting for a moment. "Maybe we are more alike than we appear to be."

Other Ben smiled. "We were raised by someone who believed in us. She taught us to be strong, albeit in different ways, which is something I will always be grateful for."

"Your doppleganger isn't wrong." Alex took Other Ben's other hand. "Mom did teach us all to be strong."

Laughing, Other Ben shook his head. "I like you, Alex. I'm sorry we won't have much time to get to know each other. Close your eyes. Keep hold of me, and let me do the heavy lifting for now."

Closing his eyes Ben let his other self take them through time, jumping from one to the next fast, something he'd only done before once, when Mom was teaching him how to bounce. It wasn't something she ever thought he'd have to do, but she wanted him to be ready for every eventuality.

"We're there." Ben opened at his eyes Other Ben's voice. "This is a time Marie hasn't changed yet, so we'll be able to find her here soon enough."

"Hopefully." Alex took a step to one side. "I'd forgotten how much I hated bouncing." He shook his head. "Give me a few seconds, and I'll be fine."

Moving to join his brother, Ben looked into Alex's eyes. "Ground."

"I know what I'm doing."

"Still, ground." Ben smiled. "Feel your roots."

"Even when Mom did it I hated that, and it's no better when you're doing the annoying part. Stop. I know what I'm doing."

"Oh, you do, do you? Like I know that. We've never jumped together before."

"Let's just say we had the same first teacher, and Mom taught us both well."

"Worked, didn't it?"

Blinking, Alex stared at Ben for a moment, and then he nodded. "She really did teach us both well. Her old technique. Annoy me until I was grounded." He shook his head. "I hate you."

"Good." They both looked at Other Ben. "Where do we need to be?"

"Follow me." Other Ben led them along the road. "I often do similar things with my little brothers." He glanced back at them. "I'm glad you're working to have a different kind of relationship. Being alone… it's hard. They help me when I'm feeling the weight more than I should be."

"It's a heavy weight to have on your shoulders, as an adopted Austin."

"Constantine wants me to be the one to take over, even though I wasn't born an Austin. He's said I'm more than capable of leading people, but I don't know if I should, because I'm not family. Not the way my brothers are."

"He chose you, and that's what matters the most. Marie, being the daughter of Samuel Austin, made the decision to go on a time rampage, causing so many splinters we might not be able to put them all back together. Blood doesn't mean anything when it comes to who the best leader is. We were both brought up by Mom, and we know what she believes. She was always going to make similar choices to Samuel, which is what makes us more similar to an Austin than Marie was, because she didn't walk that path."

They came up on a turn, and Other Ben led them down a street. Not knowing when they were did have an effect, but the upside of being with him meant their clothes were already changed, so they were wearing exactly what they needed to be to fit in. Ben looked around slowly, trying to see Marie.

Spotting her, in the midst of a crowd, Ben started moving toward her, focusing on what he might be able to grab. At the same time he slipped a hand into his pocket, wrapping a hand around the handle of the knife he had within there. It was something Mom insisted he carry, just in case, and for the first time it might actually come in handy.

Reaching her, seeing she had no idea who he was, Ben reached out to take her scarf, pulling the knife out at the same time. Their eyes met for a moment. Her eyebrows lifted, as he managed to pull her scarf from around her neck, before side stepping the blade. As the crowd reacted she shook her head.

"Katherine sending her child to do her dirty work?" Marie laughed. "Do you really think you're strong enough to do what needs to be done?"

"Yes, because I am my mother's son." Ben smiled, stepping closer to her, not caring about the screams within the crowd. "Unfortunately you aren't your father's daughter. If you were you'd know better than to create splinters the way you are."

"Dad was boring. He couldn't see what we were capable of. Once I've cleaned up history I'll be able to pull the splinters back together, before taking control of the travellers. Maybe, if you're really lucky, I might give you a chance to prove yourself to me. Katherine, however, will pay for her choice, because that is something I'm not willing to forgive."

Before Ben had a chance to say anything more she was gone. Fortunately Other Ben and Alex appeared seconds later. Ben passed Alex the scarf, so he could grip the knife, and then they were going, bouncing through time the same way Marie was. Ben knew he couldn't close his eyes. He had to keep them open, so he was ready for any chance.

Every so often Marie glanced back, to see if they were still there, the frustration in her face obvious. She didn't want to stop to grab the scarf, even though that was probably what she should do. It was the only way she'd be able to stop them. She kept going instead, always a couple of seconds ahead, making it impossible for Ben to make a move.

They kept bouncing. Alex didn't look at all happy about it, but he also wasn't going to stop them. Maybe it was something Marie noticed, because finally she did stop, moving to try to grab Alex. Ben was there before she could, not letting her touch him. She did get a hand on the scarf, almost enough to pull it away. Only Alex was better than that. He kept his grip on it, without Ben needing to help.

While they fought with Marie for the scarf Other Ben moved forward, with his own knife in his hand. He moved forward, only managing to slice down her arm before she jumped again, meaning they had to keep following her, bouncing from one time to the next. It definitely wasn't the most interesting way to see how time was different in that splinter.

Finally they reached the traveller's headquarters. Marie didn't stop. She sprinted through the hallways, because she knew when they were in a way they didn't. Reaching one of the gathering rooms it was possible for them to see what she was planning. If she removed Katherine it would take out the problem. Only it wasn't going to be as simple as the thought it was.

Other Ben moved fast. It was all Ben could do to keep up with him, both of them with knives in hand, ready to do whatever needed to be done. Before they managed to reach Marie she was in front of Katherine, who was much younger. Twenty years, however, didn't appear to have changed her. Even though it must have been a shock to see an older Marie right in front of her Katherine was ready with a punch. It dropped Marie to the ground.

As both the Bens reached Marie on the ground she bounced again, not seeming bothered by the position. They couldn't go straight away, because Alex wasn't with them. He wasn't anywhere near then. The Bens shared a look. Ben went back the way they came as Other Ben started to explain to Katherine what was going on.

Reaching the main hallway Ben stilled. There was no sigh of Alex, or the scarf. Marie took him. Biting down hard on his lip Ben ran back through the hallways, to where Katherine was with Other Ben, studying him. It was to be expected, considering the story she'd just be told.

"He's gone."

Raking a hand through his hair Other Ben pulled the list out of his pocket. "I don't even know if this is going to be any use. She could be anywhen right now, with Alex."

Katherine looked between the two of them. "Alex is my other son, right? From a different splinter?"

"Yeah, he is."

"Okay." She gestured for them to follow her. "Marie has stuff here still, even though she made the decision she couldn't stay. With what happened... well, it's understandable. She made choices Samuel was never going to accept from his daughter, and now she has to live with it, but she isn't comfortable around him any more. She thinks he should have forgiven her. It was nothing more than academic interest. Apparently."

"It wasn't."

"No, I know, and Samuel does as well. He is willing to give her another chance, if she proves herself. She doesn't believe she should have to."

They went down a hallway to where the bedrooms were. "She's not going to change."

"Samuel knows, but he wants to believe she might prove otherwise. Marie means a lot to him. Only now..." Katherine raked a hand through her hair. "I know I can't tell him, which is something I hate. He can't know what ends up happening. With what she's doing it's possible he might learn sooner rather than later. Doing what she is... well, it's always going to have an effect.

"Bouncing from time to time, creating splinters. She knows better, especially as there's a chance this can't be undone, because splintering time on purpose... it's far more complicated than doing it accidentally. Yet she still believes she's going to be able to pull them together, to create the time she wants, rather than understanding the damage she's caused."

Katherine stopped at a door, opening it to let the two of them in. Ben went to one of the drawers, pulling out another scarf, almost identical to the other apart from colour. He wrapped it around his neck, wanting to be ready when they finally found Marie. Taking Alex wasn't something he was going to forgive.

Leaving again he looked at Katherine. "Thank you."

"Save him. I don't want to grieve a son I never even knew."

"We will." Ben looked at Other Ben. "Let's go. She's got a relatively long head start. I don't know to think about what she might have done."

Nodding, Other Ben took Ben's arm. "We'll find him. I promise."

Finding him was one thing. With one last look at Katherine Ben smiled. "Good luck."

CHAPTER 9

Marie looked at the Bens. "You made it then. I wasn't sure you would. Likelihood was I'd chosen the right time, but there was no way of knowing for certain." She studied them. "Even if I hadn't I assumed you'd be able to think of something. The two of you are Katherine's sons." She looked between them. "Two of you is very interesting. Choosing to travel splinters… it makes you useful. This would all be much simpler with one of you by my side."

"No." Ben's voice was flat. "Where's my brother?"

"How about you tell me which one of you is which? Then I'll hand over your brother. Maybe." Marie smiled. "Or maybe I'll keep him, to make certain you behave. I know you want to kill me. You were taught by your mother time is sacrosanct, right? How no one should do what I have. Dad was the same. He would never forgive me for this, but that doesn't matter, because I will never forgive him either. Seeing the choices he was willing to make... well, I guess he believed he was making the right decisions. For time.

"People didn't matter to him. Not the way they did to me. I wanted to make things better, to save lives, while he told me a hundred times over some people had to die in horrible ways in order to protect time. Only in the end he wasn't able to protect time. He died, like everyone else, without changing anything.

"Yes, he gathered the travellers together, believing we should be the ones in control, but that... he was making a huge mistake. He was never meant to be in control, whatever he thought. He saw the world in black and white. To him there were no shades of grey. I am grey. Accepting me for who I am was an impossibility, and then I found Katherine.

"My best friend. The woman I shared all my secrets with, because I trusted her, until the time came when I found myself trapped in a paradox. She changed a couple of numbers, and there I was. Only it gave me the time I needed to work through a lot of things. To plan how I was going to make her pay.

"Then Edward found me. Told me Constantine was looking for me. I talked to him for a bit, he told me the choices he'd made, and then I made certain he'd never be able to get home to her, because she wasn't going to build a life when she took that away from me. Only she still had her two sons. Two sons. You're here for your brother, right?"

"Are you really doing this? I honestly thought better of you, Marie. Instead you have to start talking. Telling me all about your life, as though I care. I don't." Ben stepped closer. "Where's Alex?"

"Like I said I need one of you. I don't need the other one, and making Katherine pay multiple times over is definitely what I wanted to do." Marie looked between the two of them. "You're the one I can control with your brother, but is it really worth it? Two of you seems like a lot to deal with."

"Good." Alex was suddenly there, with a scarf around Marie's neck. "Do you really think Mom didn't teach me how to get out of something like that?"

Both the Ben's moved at the same time. Marie tried to shake Alex off, but he was angry. Using him a bait was easily the stupidest decision Marie could have made. It was only ever going to lead to him reacting badly, to him attempting to kill her, because that was what Mom prepared them for. To do what needed to be done, whatever the situation was, as she wasn't always going to be there.

Somehow Marie did manage to get Alex away from her, flinging him across the room, where he hit the wall hard. Using her own anger against them. Breathing in deeply, knowing they needed to move carefully, Ben tightened his grip on the handle. Moving forward, he prepared for any move she might make.

Other Ben went a different direction. Had he been taught the same lessons? It seemed likely, as they tried to pincer Marie between them, because it was impossible for her to see both of them at the same time. She looked between the two of them. There was no way for her to stop both of them. She had to pick one of them to go against.

Hoping it would be him Ben kept moving, keeping enough distance between then she couldn't do anything to harm him. If she managed it it wouldn't be due to him making a mistake. It would be because she was better. Something that might actually be possible, depending on what she'd done during her time in trapped in the paradox, but if she was he'd fight for his life.

Instead of going for Ben she went for Other Ben. Marie moved fast, and then Alex was there, still using the scarf as a weapon, wrapping it around her neck before she go anywhere near Other Ben. They were both protecting their not-brother. Making use of the distraction Ben moved forward, knife slipping between her ribs.

It was enough. Marie dropped to the ground, staring up at Ben, the shock in her eyes obvious. She didn't say anything, as she fought to breathe. It wasn't going to happen. Other Ben stepped closer, the three of them watching until she was dead, her eyes still staring up at them. Kneeling, Ben reached out, and she moved at the same time, hand wrapping around his neck.

Before Marie had a chance to actually do anything Other Ben shoved his knife into Marie's heart. At least he seemed to go straight into their heart, as she coughed up blood straight away, and then she was dead, her hand loosening. Ben reached out with a shaking hand to check for a pulse.

"She's really dead this time." Ben breathed in deeply. "What do we do next?"

"We should bury her." Alex moved forward. "Even though she tried to kill her it's the right thing to do."

Nodding, Ben stood, letting Alex help him stay upright. They shared a look. Killing in the abstract was easy. Actually doing it was something else entirely, but it had to be done. It was Other Ben who lifted Marie's body. As they didn't have a shovel it wasn't going to be easy to bury her, unless they took her back to the time they all come from.

"Constantine should bury her." Other Ben breathed in deeply. "Put a hand on my shoulders, and I'll take you back to when we came from. Then they can help us do what needs to be done."

Moving to him, putting their hands on his shoulder, it was obvious he was right. Constantine, even if he wasn't the right one, deserved a chance to say goodbye, and to decide what happened to her. Travelling through time was different then. Going back to their present. As they appeared, Constantine and Katherine in front of them, they knew they still had so much work to do.

Other Ben carefully put Marie's body onto the table. "It's done." He stepped back, and it was then Ben noticed he was also shaking. "I'm sorry, Constantine."

"Don't apologise, Ben. This is what needed to be done." Constantine looked at Marie's body, the pain in his eyes obvious. "Thank you for bringing her home. I can bury her in the Austin graveyard."

"Can we help?" Ben raked a hand through his hair. "This isn't something I want you to do alone."

"No, I think I need to do this alone. My goodbye to her is something I've been waiting on for a long time, and I know this isn't my sister, but this is one version of her, who wanted to change the world the same way." Constantine carefully picked up Marie. "This isn't on you. I wish you hadn't needed to be the ones to end this. I wish it could have been a different end. I'm so sorry you had to take their weight on your shoulders."

"We did what needed to be done, Constantine." Ben's eyes met with his for a moment. "It was the right choice. As hard as it was we made the right choice."

Constantine nodded, as he walked out with Marie in his arms. Ben sat down opposite Katherine. Her eyes met with his for a moment, and she put her hand out to touch his for a second, before she stood, going around the table to hug Other Ben. Alex sat down beside Ben.

"That was one strange bonding activity."

Ben shook his head, before resting it on Alex's shoulder. "Yeah, it really was, and I don't want to do it again."

"Hopefully we can make that happen, but it depends on what we manage to do with the splinters. That's the next step."

"I know." Ben breathed in deeply. "Our lives will never be the same again, will they?"

"Do we want them to be? Change isn't a bad thing, Ben, however hard it might be to go through. We made a complicated choice, because if we didn't time would have been destroyed. This way we do have a chance to fix the damage she did."

"Maybe." Sighing, Ben raked a hand through his hair. "I don't know. I think this is going to take some time to work through."

"Of course it is. If you were doing okay with this I'd be more worried. What we did today always had to change us." Alex wrapped an arm around Ben's shoulders. "Whatever happens we're going to be able to move forward, to live a better life, and we'll help to make the travellers better than they were before. We're going to stop this from happening again. No one will have to go through what we did."

"All you can do is your best." Katherine stepped over to them. "You have very high expectations of yourself, and it's something we've tried to do. Only sometimes people slip through the net. We did everything we could to protect time from those people, and we will keep going it, but sometimes everything you can do isn't enough.

"How you're feeling right now is normal. Working through all of this will take time, Ben, but you made the best choice you could, and you will get through this. Both of you are strong. I can see it. Your mom will help you through all of this when you get back, because she'll wish, the same way I do, she could have taken this weight on her shoulders. Instead it had to be you. You were the ones who could get here."

Nodding, Ben raked a hand through his hair. "Thank you."

"No, thank you." She kissed the top of his head. "You might not be my actual son, but you are my soul son, and that's enough." She did the same to Alex. "We worked while you were gone on the splinters, to see what we could do to pout them back together. Even though we aren't done yet we are getting there. It's a slow process."

"Slow processes are something we can help with." Ben looked at Katherine. "How about you get what you have, and we can continue on."

"Okay."

As Other Ben sat down with them Katherine put the papers closer to them, making it possible for the three of them to work through what still needed to be done. Each splinter needed to go back together in the right place, otherwise they were going to end up with their timeline being different to how it should be. Only there were a lot of splinters to put back together. Normally it would be three or four at a time. With what Marie did it was hundreds.

CHAPTER 10

Moving the next splinter back into place Ben went through the list again, checking what they'd already done. Twenty splinters down so far, with things going back the way they had been before, which was a good sign. There was still a chance they might make mistakes as they went through, but mistakes happened when there were so many of them to deal with.

Jumping back to Constantine and Katherine Ben waiting for Alex and Other Ben to move their splinters back into place, before they moved on to the next three. With every step they took it was one closer to their lives being back to how they once were, while also being entirely different, because there would be no Other Ben. No Constantine and Katherine who'd been together since Ben was a baby.

"It's okay. We made the choice."

"Ben didn't."

"Yes, he did, because he knows how important time is. This is something you're never going to forget, which is all we can ask for. We'll always be a part of your lives, one way or another, and that's enough for us. I promise you that." Katherine pulled Ben into a hug. "We know we can't be here. Our splinter isn't meant to be. As hard as it is to say goodbye we don't have any other option. Not if we want to do things right."

"Right is the hardest choice."

"Of course it is. Marie believed she was doing what was right, however many times she was told she'd end up destroying the timeline, so right isn't always what it seems to be. It's when Samuel created the travellers in the first place. Had he not made that choice we wouldn't be where we are now."

"We wouldn't, and Father is the reason we'll keep doing what needs to be done. That doesn't mean what happened won't affect your present in some way. That's what often happens in this kind of complicated situation." Constantine shrugged. "Be ready for that when you go back."

Nodding, Ben smiled. "I'm as ready for it as I can be."

Alex returned next, then Other Ben, and then they moved on to the next splinters. Slowly, one by one, they put time back together, until the last splinters remained. Ben looked at Other Ben, their eyes meeting for a moment. Looking down, breathing in deeply, Ben tried to find the words he needed, but they weren't there. How did he say goodbye to his other self? It didn't seem possible.

"Everything is going to be fine. This is how things are meant to be."

"Knowing that doesn't make it any easier. You were by my side through one of the hardest times of my life. That's... I don't know how I'm supposed to do this."

"Which is exactly the reason we shouldn't create splinters. We get attached to the people within them. You have to say goodbye, Ben, and that's not ever going to be easy. It's what Mom taught us. We make the right choices, because if we don't then no one else will. So you say what you have to, so you can move forward, otherwise we're going to fail those who need us."

"Literally goodbye, Ben?" Ben smiled, even though it wasn't the easiest thing to do. "Thank you, for everything, because we wouldn't have been able to do this without you."

"You're welcome." Other Ben smiled back, but there was still a mix of emotions in his eyes, whatever he might be saying. "Good luck with the future, and make certain you walk the path we know is right."

"I will. Goodbye, Ben."

"Goodbye, Ben."

Travelling to the last splinter, leaving Other Ben to do his splinter, Ben waited to feel the change. It was something he connected with so much better than he had done before, after putting together a hundred splinters of time. As it happened he knew his other self was gone, the two of them to never meet again, unless something untoward happened again. Hopefully it wouldn't.

Another of Alex's splinters went in to place, and then Ben gently urged the last one back to where it was meant to be, taking him back to the splinter he'd left behind hours earlier. Even though it felt like days it wasn't. Constantine was waiting for them. As their eyes met Ben was able to see the acceptance in his.

"Well done. You did good work." Constantine stood. "Peter believes you're going to be exactly the kind of traveller we need right now."

"Constantine..."

"Of course you're going to say you shouldn't, because you're the son of Katherine and Edward, but I think things are going to change for the better soon enough."

"Did Alex get back okay?"

"Yeah, he's in the library. Apparently putting splinters back is hungry work."

"In a way it is."

"You aren't ready."

Even though it wasn't a question Ben nodded. "This has been a lot."

"Something I've always been good at is listening to people, if you want to talk about what it was like. I understand if you don't. What you did today is something most people wouldn't have to, and yet you did it because it was what was right, however hard it ended up being."

"Killing someone is a choice I never want to have to make again."

"Good." Constantine put a hand on Ben's shoulder. "I can't imagine what you must be feeling right now."

"We ended up in a splinter where you and Mom got together after what happened with Dad, and my other self..." Ben breathed in deeply, looking down at the ground. "It almost feels like I killed them too. This was what they wanted, knowing what we do, but they don't exist any more."

"Somewhere they do, at least in theory."

Ben raised an eyebrow. "Multiverse theory."

"I don't know if I believe it, but if it helps then maybe you should."

"Everything... I wish I knew what I was supposed to do next."

"No one ever knows what they're supposed to do next after something like this is over. Alex wanted to eat, which is understandable. You don't seem to want to do anything."

"Alex always wants to eat. At least it seems like it." Ben smiled. "I'm glad he's doing okay."

"Come on. I know Katherine wants to see you, make sure you're whole, and I think it would help to ground you back in this time."

Nodding, Ben followed Constantine to the library, where Alex sat with a sandwich in front of him. "There you are." Alex looked at Ben. "How are you?"

Shrugging, Ben sat down opposite Alex. "I said my goodbyes, and now…"

"I know. I did the same. It's complicated, but we both know this is what we were supposed to do. Pain, sometimes, is a necessity."

"Unfortunately. I've experience it." Mom took the seat next to Ben, wrapping an arm around his shoulders. "I'm sorry. This should never have been your problem to solve. I should have found a better way before."

"Do you really think there was one?"

"Maybe not, but I feel like I should have tried harder. Marie... she was my friend. What happened... Alex told me she was angry with me for what I did, and wanted to use you to create the kind of world she believed this should be."

"Yeah, she did, and she took Alex to try to make it happen."

"That was a very stupid decision to make."

Laughing, Alex nodded. "She came to regret it." The amusement faded. "It was the right choice, but it was hard. Doing what we did... it's something we'll both live with for the rest of our lives."

"Of course you will. That's because you're normal, and aren't willing to harm for no reason. What you did was necessary. Horrible, but necessary."

"What happens next?"

"Now that's the more complicated question to answer." Mom looked between the two of them. "First I need to free Edward, and then we need to have a long conversation about what our is going to happen with the travellers."

Constantine sat down next to Alex. "The reason we couldn't make it work before was due to me. I wasn't entirely willing to put my foot down, even though I should have done. What happened today is a good reason for us to force certain rules on the travellers, if they're going to keep travelling.

"Peter is looking into a way we're going to be able to stop them from having the freedom to travel permanently. It may not be possible, but if we can stop this from happening again then we should, and paradoxes aren't a safe way to stop them. As we've seen with Marie it is possible for someone to get out, if they're very determined.

"Fortunately we do have ways of dealing with that problem if it happens. We'd rather it didn't, though, so the focus needs to be on prevention, which will definitely mean more lessons before someone is permitted to travel. I'm going to be taking back the position of leader, because it's past time there was an Austin in charge.

"At the same time I know I need someone to take over in the future. Ben, I want that to be you, with Alex by you side if that's the decision you make together. The two of you have already done so much for us, so I know it's a lot to ask. You can say no. This is something I won't force you into, if you don't believe it's something you can do."

Ben and Alex looked at each other across the table. Alex shrugged. "Three days ago I thought Alex would be willing to shove me into a paradox if I dared do anything like this, in an attempt to save Edward. I'm not certain I can wrap my head around how everything has changed. It's going to take time."

"Yes, it will, and this is something you don't have to decide or right now. I just want you to know my plans. The two of you have been brought up by a woman who should have been an Austin, so you're going to make the right choices."

"We would do the best we can." Alex smiled. "A very smart woman told us not to put high expectations on our shoulders. That we should accept sometimes bad things are going to happen however many good decisions we make, which is what we saw with Marie. She was brought up by an Austin, and yet we still found ourselves in this position, because sometimes it's not possible to stop them from happening."

"A very smart woman?" Mom looked between the two of them. "Are you talking about me?"

"I am, in a different splinter, and she was like you, except the travellers were the ones who were pushing the rules. Others worked against them, so we know it's possible for us to be able to make this happen the way you're seeing now. I can't help wonder if that splinter is having an effect on this one, as it was the closest one to ours. It was where Edward come from, before he made the journey here to try to get Mom to give him a chance."

"That was never going to happen."

"No, but he didn't know that." Ben shook his head. "His Katherine chose to move on with Constantine, and it was something Edward couldn't accept, so he travelled through splinters to find us. It wasn't because he wanted to help us. It was because he wanted you, Mom, believing it would be another chance."

"Like I said when he brought it up I'm not going to get into a relationship with someone who's now twenty years my junior. I've lived a whole life without him. I've brought up two amazing sons, who have helped to save time, and I can't imagine being prouder of the time of you."

"We did what had to be done, nothing more." Ben looked at Alex, who nodded. "Hopefully tomorrow will be an easier day."